SCIENTIFIC AMERICAN EDUCATIONAL PUBLISHING

BODY SCIENCE

10 FUN EXPERIMENTS ABOUT THE HUMAN BODY

BRING SCIENCE HOME

Published in 2023 by Scientific American Educational Publishing in association with
The Rosen Publishing Group 29 East 21st Street, New York, NY 10010

Contains material from Scientific American, a division of Springer Nature America, Inc., reprinted by permission as well as original material from The Rosen Publishing Group.

First Edition

Editor: Therese M. Shea
Designer: Rachel Rising

Activity on p. 5 by Science Buddies/Svenja Lohner (August 15, 2019); p. 11 by Science Buddies (February 13, 2014); p. 17 by Science Buddies/Sabine De Brabandere (May 16, 2019); p. 23 by Science Buddies/Sabine De Brabandere (October 4, 2018); p. 27 by Science Buddies/Svenja Lohner (April 27, 2017); p. 33 by Science Buddies/Megan Arnett (March 9, 2017); p. 41 by Science Buddies (November 19, 2015); p. 47 by Science Buddies (December 27, 2012); p. 52 by Science Buddies/Daisy Yuhas (May 31, 2012); p. 56 by Science Buddies (March 8, 2012).

Photo Credits: pp. 3, 5, 8, 9, 11, 14, 15, 17, 21, 26, 30, 37, 38, 45, 50, 54, 56, 59 cve iv/Shutterstock.com; pp. 4, 5, 11, 17, 23, 27, 33, 41, 47, 52, 56 Anna Frajtova/Shutterstock.com.

All illustrations by Continuum Content Solutions

Library of Congress Cataloging-in-Publication Data
Names: Scientific American, inc., issuing body, editor.
Title: Body science : 10 fun experiments about the human body / Scientific American.
Description: Buffalo, NY : Scientific American Educational Publishing, an imprint of Rosen Publishing, [2023] | Series: Bring science home | Includes index.
Identifiers: LCCN 2022039847 (print) | LCCN 2022039848 (ebook) | ISBN 9781725348899 (library binding) | ISBN 9781725348882 (paperback) | ISBN 9781725348905 (ebook)
Subjects: LCSH: Human body--Experiments--Juvenile literature. | Human body--Juvenile literature. | Human physiology--Experiments--Juvenile literature. | Human physiology--Juvenile literature. | Science projects--Juvenile literature. | Science--Experiments--Juvenile literature.
Classification: LCC QP37 .B677 2023 (print) | LCC QP37 (ebook) | DDC 612--dc23/eng/20220826
LC record available at https://lccn.loc.gov/2022039847
LC ebook record available at https://lccn.loc.gov/2022039848

Manufactured in the United States of America

CPSIA Compliance Information: Batch #CWRYA23. For further information contact Rosen Publishing, New York, New York at 1-800-237-9932.

CONTENTS

⚛ THESE ACTIVITIES INCLUDE SCIENCE FAIR PROJECT IDEAS.

INTRODUCTION

The human body is beyond amazing. So much is happening within your body every second of your life—and you don't even need to think about most of it. To hone an appreciation for what's under your skin—and your skin itself—it's helpful to learn more about bodily functions. To do that, hands-on experiments like the ones in this book will help you!

Projects marked with ⚛ include a section called Science Fair Project ideas. These ideas can help you develop your own original science fair project. Science fair judges tend to reward creative thought and imagination, and it helps if you are really interested in your project. You will also need to follow the scientific method. See page 61 for more information about that.

Build a Blood-Flow Model

GOING WITH THE FLOW: LEARN HOW OUR BODIES MOVE BLOOD BEST BY BUILDING A MODEL OF THE BODY'S CIRCULATORY SYSTEM.

Why is it important to eat well and get exercise? Doing these things keeps your blood flowing well through your circulatory system, which is key to staying healthy. The job of your circulation is to transport blood through your heart, veins, and arteries to provide oxygen and nutrients to your body. If this blood flow doesn't work properly, it can negatively impact health. In this activity, you will find out what happens to the blood flow of people whose circulatory systems are impacted by a condition called coronary heart disease.

PROJECT TIME

30–45 minutes

KEY CONCEPTS

Biology
Health
Medicine
Circulatory system

BACKGROUND

Every part of the human body needs to be supplied with oxygen and nutrients. This task is accomplished by the body's circulatory system, which is also called the cardiovascular system (and includes the heart, veins, and arteries). Blood transports oxygen and nutrients by traveling through the arteries and delivering them to the other parts of the body. The heart acts as a powerful pump that generates the force necessary to move the blood around the circulatory system. The arteries carry blood from the heart to the lungs, where it is oxygenated, and then to the organs. Veins return the blood to the heart. When something goes wrong with the body's circulatory system, it can lead to serious health consequences.

The most common problem with the circulatory system today is coronary heart disease (also called coronary artery disease). Coronary heart disease is caused by the buildup of cholesterol and other material, called plaque, in the arteries. Over time, the plaque grows thicker, and the arteries become narrower. As the arteries narrow, they cannot carry as much blood to the organs. The decrease in oxygenated blood can lead to many health problems.

In this activity, you will model what happens to blood flow when coronary heart disease narrows a person's arteries. You will do this by comparing the blood flow through straws (arteries) of different diameters. Of course you won't use real blood but colored water instead!

MATERIALS

- Two 16-ounce (473 ml) disposable plastic cups
- Two straws with different diameters
- Modeling clay
- Stopwatch
- Tray or shallow tub (large enough to hold more than 16 ounces [473 ml] of water)
- Water
- Red food coloring
- Measuring cup (with ounces and milliliters)
- Scissors
- Permanent marker
- Paper
- A workspace that can tolerate spills
- Adult helper

PREPARATION

- Measure 13.5 ounces (400 ml) of water, and pour it into the first plastic cup. Mark the water level on the outside of the cup using a permanent marker. Then make a mark on the other cup at the same height. Now pour the water out of the cup so it's empty.

- Have an adult help to carefully use the scissors to poke a hole into the side of each cup about 1 inch (2.5 cm) from the bottom. Widen the hole enough to fit each straw.

- Cut a 2-inch (5 cm) length from each straw. *What do the straws represent in your blood-flow model?*

- Insert the straws into the holes. Make sure they are not squeezed. The straws should point down on the outside of the cup so water can flow out easily.

- Use modeling clay to seal the inside and outside of the hole around the straw.

- Prepare 34 ounces (1 l) of water. Add red food coloring to the water. *What does the red water represent in your blood-flow model?*

PROCEDURE

- Start by placing the cup with the wider straw into the tray. *What condition do you think the cup with the wide straw represents—a healthy or unhealthy cardiovascular system? Can you explain why?*

- Seal the end of the straw with a piece of modeling clay.

- Fill the cup up to the mark with red water. (Improve the clay seal if needed, and make sure you are starting with water up to the line.)

- Get your stopwatch ready! Remove the clay seal from the end of the straw, and immediately start the stopwatch. Stop the stopwatch when the water stops flowing through the straw. *How long does it take until the water stops flowing out of the cup?* Record the time on a piece of paper.
- Place the cup with the narrower straw into the tray. *What condition does the cup with the narrower straw represent? Why do you think this is the case?*
- Seal the end of the narrow straw with a piece of modeling clay.
- Fill the cup with the narrow straw up to the mark with red water. (Improve the clay seal, and add more water if needed.)
- Remove the clay seal from the end of the straw, and immediately start the stopwatch. Stop the stopwatch when the water stops flowing. Record the time on a piece of paper. *How long did it take this time until the water stopped flowing?*
- Compare both times that you recorded on your paper. *Which condition (wide or narrow straw) resulted in a faster and better blood flow? What does your result say about the blood flow in people with coronary heart disease?*

EXTRA

Calculate the flow rates of the "blood" in both of your setups (wide and narrow straw) by dividing the water volume at the beginning by the time it took for the water to flow out of the cup. *How do both numbers compare?*

SCIENCE FAIR IDEA

Instead of using two different-sized straws, use two identical straws. Then use modeling clay to "clog" one of the straws to model the plaque buildup inside an artery. *What happens to the speed of the water exiting the straws?*

OBSERVATIONS AND RESULTS

You probably guessed correctly that the straws in your blood-flow model represent the arteries in your body that transport blood from your heart to your organs. The red water in your model is the blood that flows through the arteries. Testing the setup with the wide straw simulates a healthy cardiovascular system in which the arteries are free from any plaque. As you likely noticed, the "blood flow" out of the cup should have been relatively fast. The second setup with the narrow straw represents an unhealthy cardiovascular system. The artery (straw) is starting to get clogged with plaque and thus the diameter of the artery gets smaller and smaller. This also means that less blood can flow within the artery, which you should have observed in the activity. With the narrower straw, the water takes much longer to flow out of the cup. In the worst cases of coronary artery disease, the whole artery can get clogged and no blood can flow to organs anymore. This is why it's important for people to get exercise and eat healthy foods, which helps keep arteries open to good blood flow.

CLEANUP

Pour the red water into the sink. Recycle the cups if possible, and dispose of the straws in the trash. You can dry your modeling clay and reuse it.

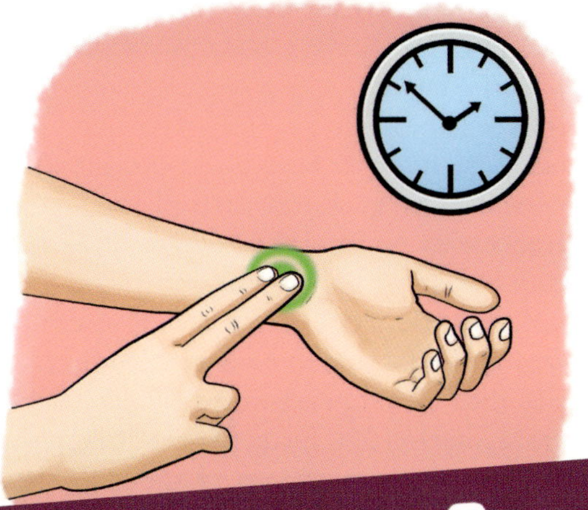

Cardiovascular System Science

Investigate Heart-Rate Recovery Time

KEEP IT MOVING: FIND OUT HOW HEART RATES CHANGE DURING AND AFTER EXERCISE AND HOW HEART RATES ARE LINKED TO HEART HEALTH.

Real hearts don't look much like the two-lobed shapes adorning cards and candy boxes. The actual shape of the human heart is important for its function of supplying blood to the entire body. You have likely noticed that your heart beats more quickly when you exercise. But have you ever taken the time to observe how long it takes to return to its normal rate after you're done exercising? In this science activity, you'll get to explore more about your own or another person's heart-rate recovery time.

PROJECT TIME

30-45 minutes

KEY CONCEPTS

Heart rate
Exercising
The heart
Cardiovascular system
Health

BACKGROUND

Your heart is continuously beating to keep blood circulating throughout your body. Its rate changes depending on your activity level; it is lower while you are asleep and at rest and higher while you exercise—to supply your muscles with enough freshly oxygenated blood to keep them functioning at a high level. Because your heart is also a muscle, exercise, in turn, helps keep it healthy. The American Heart Association recommends that a person does exercise that is vigorous enough to raise their heart rate to their target heart-rate zone—50 percent to 85 percent of their maximum heart rate, which is 220 beats per minute (bpm) minus their age for adults—for at least 30 minutes on most days, or about 150 minutes a week in total. So for a 20-year-old, the maximum heart rate would be 200 bpm, with a target heart-rate zone of 100 to 170 bpm. (For those 19 or younger, target zones can vary more than they do for adults.)

After exercising, a person's heart needs time to recover, or to return to its normal resting heart rate. How long it takes for the heart to resume its resting rate is referred to as heart-rate recovery time. In general, people who exercise regularly, and therefore are more likely to have healthier hearts, have faster heart-rate recovery times than people who do not regularly exercise. So after a 100-meter dash, an Olympic sprinter would return to a resting heart rate faster than someone who rarely ever runs.

MATERIALS

- Pen or pencil
- Scrap piece of paper
- Clock or stopwatch that shows seconds (You can also use the stopwatch app on a smartphone.)
- Simple and fun exercise equipment, such as a jump rope, Hula-Hoop, stepping stool, etc. (Alternatively, you can do exercise that does not require equipment, such as jogging or jumping jacks. Always stop exercising if you feel faint or light-headed.)

- Chair
- Calculator (You can also use a smartphone app for this.)
- Adult volunteer who can safely exercise (This is optional, but will allow you to calculate more precise target heart-rate zones.)

PREPARATION

- Practice finding your (or your volunteer's) pulse. You can do this by using the first two fingers on one hand to feel the radial pulse on the opposite hand's wrist. The radial pulse is found on the "thumb side" of the wrist, just below the hand. You will need to be able to find the pulse quickly, so practice finding it until it is easy to do.

- You can alternatively use a carotid pulse, but make sure you know how to take it safely by pressing on the neck only very lightly.

- Prepare a scrap piece of paper so that you can quickly record your testing results when doing the activity. You will be measuring the resting heart rate, the heart rate immediately after doing a short exercise, and then every minute for the next five minutes. After that you will measure the heart rate every two minutes. You will continue to measure it until it returns to its resting rate.

PROCEDURE

- Measure your or your volunteer's resting heart rate. To do this, take the pulse (after several minutes of resting) and multiply the number of beats you count in 10 seconds by six. This will give you the resting heart rate in beats per minute (bpm). *What is the resting heart rate?* Write it down on your prepared scrap piece of paper.

- Choose a vigorous activity. This could be jumping rope, stepping on and off a low stool, Hula-Hooping, jogging, etc. Then, engage—or have your volunteer engage—in that activity for about two minutes.
- After the end of the exercise period, immediately sit or have your volunteer sit on the chair and quickly measure and record the heart rate (again, counting the number of beats in 10 seconds and multiplying that by six). *What is the heart rate after exercising?* Write this number down.
- Now, record what time it is (including seconds) or start a stopwatch.
- While you or your volunteer continue to sit and rest, measure and write down the heart rate every minute for the next five minutes. *How does the heart rate change as you continue to rest after exercising stopped?*
- If the heart has not reached the resting rate, continue to rest, but only measure and record the heart rate every two minutes. Continue measuring it until it returns to the resting rate. *How does the heart rate continue to change over time?*
- *What was the heart-rate recovery time, or the amount of time it took the heart to return to its resting rate after exercising? How did heart rate change over time after exercising stopped?*

EXTRA

Graph your results. How did heart rate change over the course of this activity?

EXTRA

Do this activity again, but this time, include volunteers of different ages. (Be sure that the volunteers you recruit can safely exercise.) *Does heart-rate recovery time change with age?*

SCIENCE FAIR IDEA

Try repeating this activity, but compare a group of regular exercisers with a group of nonathletes, or people who have a more sedentary lifestyle. *How does the heart-rate recovery time of people who are physically fit compare with people who do not exercise regularly?*

SCIENCE FAIR IDEA

Design an activity to compare the heart-rate recovery time of people who smoke with people who don't. *Do smokers have increased heart-rate recovery times compared with nonsmokers?*

OBSERVATIONS AND RESULTS

Did the heart rate quickly drop after exercise ceased and then start to level out as it approached its resting rate? Did it reach the resting rate after one to seven or more minutes when exercising stopped?

Immediately after exercising, the heart's rate was likely in the upper end of its target heart-rate zone. As soon as resting started, its rate should have quickly decreased. Specifically, one minute after rest started, the heart rate likely rapidly dropped. After that, it likely continued to drop, but much more slowly as it approached its resting rate over the following minutes. It may have taken about one to seven or more minutes (after exercise stopped) for the heart to resume its resting rate. Generally, the faster a person's heart rate recovers, or reaches its resting rate, the better shape he or she is in.

CLEANUP

Put away the equipment you used as well as the writing tools. Recycle the paper if possible. Don't forget to thank volunteers for their help.

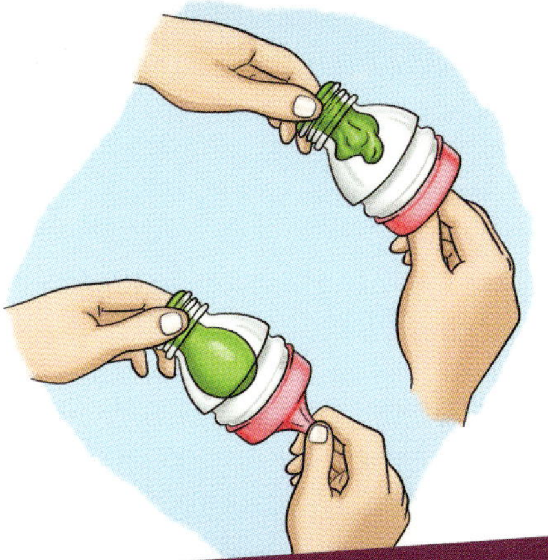

How Do We Breathe?

IN AND OUT: LEARN HOW PHYSICS PLAYS A ROLE IN EVERY BREATH YOU TAKE.

We breathe a lot—roughly 10 times a minute! Have you ever wondered how the process of breathing works so smoothly? Our lungs allow us to inhale the oxygen our body needs, but they do much, much more. They also allow us to get rid of carbon dioxide, the waste product created in the body, and they play a vital role in singing, shouting, and even giggling. In this activity, you will make a model of a lung and use it to discover how air flows in and out of the lungs with ease.

PROJECT TIME

30-45 minutes

KEY CONCEPTS

Biology
The body
Lungs
Physics
Air pressure

BACKGROUND

All cells in our body need oxygen to create energy efficiently. When the cells create energy, however, they make carbon dioxide. We get oxygen by breathing in fresh air, and we remove carbon dioxide from the body by breathing out stale air. But how does the breathing mechanism work?

Air flows in via our mouth or nose. The air then follows the windpipe, which splits first into two bronchi: one for each lung. The bronchi then split into smaller and smaller tubes that have tiny air sacs at their end called alveoli. We have millions of alveoli in our lungs! These sacs have thin walls—so thin that oxygen and carbon dioxide can pass through them and enter or leave our blood. The blood transports oxygen to almost every part of the body. The blood also gives the carbon dioxide a ride back to the lungs.

Lungs take up most of the space in the chest. The 12 pairs of ribs in our ribcage protect the lungs and other organs in our chest cavity, such as our heart.

Relaxed breathing is a reflex; we do not have to think to breathe. During this unforced inhalation, our diaphragm—the dome-shaped muscle between the chest and the abdominal cavity—flattens. This expands the chest cavity, and as a result, air is drawn in. During exhalation the diaphragm relaxes, the lungs naturally recoil, and air is gently pushed out.

We can also breathe more forcefully. When we exercise, sing loudly, or otherwise need or want more air or oxygen, we can exert force to breathe more deeply. We use various muscles to increase chest volume more dramatically. In the same way as in relaxed breathing, the expansion of the chest cavity draws air in so the lungs fill up.

The relaxation of the chest cavity pushes air out. Muscles can also force the chest cavity to contract even further, pushing even more air out. Because the expansions and contractions are larger in this case, a bigger volume of air flows in and out of our lungs, and our body gets a larger supply of oxygen or we have more air to create sound.

MATERIALS

- Disposable empty transparent bottle made of hard plastic (such as a 16-ounce [473 ml] sports drink bottle)
- Ruler
- Two balloons (8-inch [20.3 cm] balloons work well)

- Utility knife (have an adult help and use caution when using the knife)
- Adult helper
- Scissors
- Drinking straw (optional)
- Modeling clay (optional)
- Tape (optional)
- Additional balloon (optional)

PREPARATION

- Ask an adult to cut the plastic bottle. Cut off the bottle's bottom so that when a balloon hangs inside the bottle from the spout there is about 1/3 to 3/4 inch (8.5 to 19 mm) of empty space below the balloon.

- Place the cut bottle down on the wide opening. Lower a balloon into the bottle until only part of the balloon's neck sticks out. Fold the neck of the balloon over the top of the bottle. The balloon represents a lung.

- Turn the bottle over (keeping the balloon inside) so the bottle top rests on the table. In the next steps, you will create and add the diaphragm to your model.

- Make a knot in the neck of the second balloon. At the opposite side of this balloon, cut off about a third of the balloon so you are left with a wide opening.

- Stretch the wide opening of the cut balloon over the wide opening of the bottle. Pull the edges of the balloon far enough up the bottle so the balloon surface is gently stretched. Make sure that the knot is on the outside and located near the middle of the bottle opening.

- Like an inflated balloon, our lungs are full of air. We have two lungs, which are enclosed in the ribcage and protected by 24 ribs. When you breathe in, air flows into your lungs. When you breathe out, air flows out of your lungs. The balloon inside the bottle is like one of your lungs. The bottle is like your ribcage.

PROCEDURE

- Hold the bottle so you can see the balloon inside (representing the lung). Gently pull down on the knot. *What happens to the balloon inside the bottle?*

- Let the knot come back to its neutral position and then gently push it in. *What happens to the balloon inside the bottle now?*

- Repeat these steps a few times. *Does this resemble breathing? Why? Which part resembles breathing in and which part resembles breathing out?*

- If your model is working well, air will rush into the balloon when you pull the knot outward and flow out when you push the knot inward. *Why do you think this happens?*

- When we breathe in a relaxed way, our diaphragm—the muscle that separates the chest cavity from the abdominal cavity—moves to expand and contract the chest cavity. *How is that similar to what you do with your model?*

- Push and pull the knot a few more times. *Using the model, can you find which movement of the diaphragm creates inhalation and which creates exhalation?*

- Feel your ribs and breathe in deeply, then exhale. *Can you feel your ribcage expand and fall back?*

- The center of our diaphragm moves more when we take deep breaths: up to 4 inches (10.2 cm)! In the model you made, the ribcage (the plastic bottle) is fixed, but you can move the "diaphragm" more by pulling the knot farther and pushing it in more. Try it out. *How does that change the volume of air that flows in and out of the lung balloon?*

EXTRA

Add a windpipe to your model. To do this, take the balloon out of the bottle and slip its neck over a straw; secure the balloon to the straw with tape. Hang the balloon—and a short section of the straw—in the bottle's neck, and use clay to hold it in place. Make sure the clay makes an airtight seal around the straw and the bottle neck. No change is needed to the second balloon that closes off the bottom of the bottle. *Can you see which part models the windpipe?*

SCIENCE FAIR IDEA

Find a way to create a model that includes a windpipe that splits into two bronchi, each with a lung attached. The model with a windpipe and one lung is a good start. *How can you add a second lung? Can you find a reason why having two lungs is beneficial for us?*

OBSERVATIONS AND RESULTS

When you pulled the knot back, the space inside the bottle increased and your balloon probably filled up with air. In the same way, when the diaphragm in our body pulls back, the chest cavity increases, air flows into our lungs, and we inhale.

When you pushed the knot in, the space inside the bottle decreased, and the balloon probably deflated. In the same way, when the diaphragm relaxes, the chest cavity decreases, air is pushed out of the lungs, and we exhale.

When you pulled and pushed the knot further, the balloon inflated and deflated more. This mirrors what happens when a bigger volume of air is displaced when we breathe more deeply.

This dynamic works because of air pressure, a measure of how hard air presses against objects. Air pressure increases when you decrease the amount of space the air has—and decreases when you give air more space. Close a flimsy empty plastic bottle and try to compress it. It is difficult! The air inside pushes back. Open the bottle, and try to compress the bottle again. It is much easier. The air presses back with a much-reduced force. Unless something blocks the movement, air will move from areas of high pressure to areas where the pressure is lower, and this is what happens when air rushes in or out of the lungs. When the chest cavity expands, there is more space around your lungs. In this condition, the lungs can expand, making it a low-pressure area, and air rushes in to balance out the difference in pressure. Then, to breathe out, the chest cavity and lungs shrink. This raises the air pressure in your lungs, and the air rushes back out.

CLEANUP

Recycle the plastic bottle pieces if possible. Dispose of balloon pieces, and put away the intact balloon and the tools, including the scissors, knife, and ruler.

Mane Science

Why Does Wet Hair Behave Differently?

WET WEIRDNESS: LEARN HOW A LITTLE PHYSICS CAN EXPLAIN WHY SOME MATERIALS BEHAVE SO DIFFERENTLY UNDER WATER.

Have you ever noticed how hair floats freely when submerged in water, but when wet and surrounded by air, it clings together? You can see this in animal fur as well: When wet dogs shake themselves, their hair clings together in strands. Try this activity to see why wet hair is far less fluffy than dry hair!

PROJECT TIME

10–20 minutes

KEY CONCEPTS

Physics
Water
Surface tension
Cohesion

BACKGROUND

Water can also be called H_2O because the smallest amount of water that is still water (called a water molecule) consists of two hydrogen (H) atoms and one oxygen (O) atom. Once the H_2O molecule is broken, you no longer have water. This molecule is tiny—an average droplet of water contains about 1.5 sextillion molecules of H_2O!

It sometimes looks like there is a thin film around a droplet of water that holds it together. This is a result of surface tension. H_2O molecules attract one another and like to hold on closely. This behavior is called cohesion. The molecules can move around alongside one another, but they stay connected. When an H_2O molecule is in the middle of a droplet, it feels attracted to all of the surrounding H_2O molecules. It is pulled in all directions equally, so there is no preferred direction to move. That is not the case, however, for an H_2O molecule located on the edge of the droplet. This molecule feels attracted to those inside the droplet but none on the outside. It feels an inward attraction, thus only moves inward, holding the droplet together.

If you look at submerged hair and wet, nonsubmerged hair—or fur—you can see the effect of cohesion and surface tension. In this activity, you will see this effect with paintbrushes.

MATERIALS

- Paintbrushes (round ones work best; if available, select brushes with different thicknesses, types of bristles, and so on)
- Bowl or sink filled with water
- Towel

PREPARATION

- Gather your materials in a work area that can get wet.

PROCEDURE

- Start with a dry paintbrush. Observe its bundle of bristles. *What shape is it? Is it fluffy or slick? Can you easily distinguish single bristles? What is holding the bristles together? How easy is it to pull one to the side?*
- In a moment, you will submerge the brush in water. *Do you expect the bundle of bristles to look different? If so, how and why do you think this happens?*
- Submerge your brush and observe the bundle when surrounded by water. *Did the shape change? Does it look fluffier or slicker? Can you distinguish single bristles as easily?*
- *How do you think an individual bristle will respond if you try to pull it to the side? Why?*
- Get your hands in the water and feel how easily you can pull one bristle to the side.
- *Was this as predicted? If you had an explanation, would you like to change it?*
- *What do you think the bundle of bristles will look like when you take it out of the water?*
- Remove your brush from the water and observe. *What shape is the bundle of bristles? Is it fluffy or slick? Can you easily distinguish single bristles? Why do you think it looks this way when wet but not submerged?*
- Try to pull one bristle to the side. *How easy is it?*
- You only tested one paintbrush. Maybe a different paintbrush will be different. Repeat the activity with another paintbrush. *Does the same thing happen?*

ExTRA

Pay attention to your own hair. *How does it move when it is still dry? How does it move when submerged in the water? How does it move when you emerge from a bath? Why would this happen?*

ExTRA

Observe the fur of animals when they are wet. *Does it also cling together when wet and exposed to air?*

ExTRA

Can you find other examples where the wet version of an item sticks together more than the dry version does? Do you think this can also be explained by surface tension?

OBSERVATIONS AND RESULTS

The bundle of bristles was probably a little fluffy when it was dry or submerged in water. You could likely easily see the single bristles and pull one to the side. But you probably could not do this with a wet bundle of bristles once it was no longer submerged.

When pulling the bristles out of water, particles of water (H_2O molecules) stick to them. That is what we call "wet." Because water is a cohesive substance, its particles like to be surrounded by other ones so they bundle up. Thanks to that behavior, they keep the individual wet bristles tightly packed. The water particles on the outside feel pulled toward water particles on the inside of the bundle of bristles. This is referred to as surface tension, and it makes it harder to pull one bristle away.

When submerged in water, the bristles are also wet, but the water particles around it can freely move in all directions. This allows the bristles to flutter out in any direction. A wet brush exposed to air dries as the water particles evaporate. Dry bristles can also freely move because they are surrounded by air. This also happens with human hair or animal fur.

CLEANUP

Empty the bowl or tub of water. Put the towel in the laur
Dry the paintbrushes and put them away for future use.

Finding Fingerprints

DUST FOR CLUES: LEARN HOW A LITTLE CHEMISTRY CAN HELP YOU UNCOVER HIDDEN FINGERPRINTS ALL AROUND YOU!

Do you know that you leave fingerprints on everything you touch? If your hands are very dirty, this is obvious because you can actually see them. But even if your hands seem clean, your fingerprints will stay behind on the surfaces you touch—they are just invisible! Do you want proof? Then make them visible in this activity and collect your own fingerprints!

PROJECT TIME

30-45 minutes

KEY CONCEPTS

Forensic science
Chemistry
Biology

BACKGROUND

Detecting invisible fingerprints is an important task in forensic science, a branch of science that helps criminal investigations by collecting and analyzing evidence from crime scenes. Each fingerprint pattern is unique to a specific person, and therefore a reliable way of identifying a suspect. There are different types of fingerprints that can be left behind: an imprint in a soft surface, such as wax or soap; a patent fingerprint, visible to the naked eye, such as those resulting from dirty hands; and latent fingerprints, which are invisible.

These invisible latent fingerprints are made of water, fatty acids, amino acids, and triglycerides—in other words, they result from the oil and sweat that your skin produces naturally. To make them visible, you have to find a way to detect one of these substances present in the invisible fingerprint. The easiest method is called dusting, in which you use a very fine powder that can stick to the oil in the fingerprint. Once the fingerprint becomes visible, you can lift it from the surface with clear tape and transfer it to another surface to then take into the laboratory to analyze further. Other methods include using chemicals that react with the amino acids or water in the fingerprint; the chemical reaction results in a colored fingerprint, which you can then analyze easily.

Many factors determine the quality of a fingerprint on a surface. One of the most important factors is the surface texture. Fingerprints are most easily detected on smooth, nontextured dry surfaces. The rougher or more porous the material the more difficult it will be to get good fingerprint evidence. Another factor is the skin condition of your fingertips. If they are very sweaty and oily, you are more likely to leave behind prints than if they are dry and clean. Test it yourself, and collect your own fingerprint evidence like a real crime scene investigator in this activity!

MATERIALS

- A glass or smooth metal surface (where you have permission to test for fingerprints!)
- Raw cacao powder if your test surface is light; baby powder if it is dark
- White paper if your test surface is light; black paper if it is dark
- A small bowl
- Clear tape
- A fine brush with soft bristles, such as a makeup brush

- Dust cloth
- Water
- Soap
- Hand lotion

PREPARATION

- Pour a little cacao or baby powder into a small bowl.
- Wipe the smooth glass or metal surface that you are using for your experiment thoroughly with the dust cloth.

PROCEDURE

- Choose one finger of one of your hands and press it firmly onto the clean, smooth surface. Remember where you put it on the surface. *Can you see your fingerprint?*
- Dip your brush carefully into the cacao or baby powder. (If your fingerprint is on a dark surface, use the baby powder; if on a light one, cacao.) Only the tips of the bristles should be covered with powder. Then, tap the brush to remove any excess powder.
- Gently sweep the brush carefully over the area where you put your fingerprint. Make sure to not apply too much pressure so that you do not wipe the print away. If necessary, add more powder to the brush, but be careful not to add too much. *What happens when you apply the powder to the surface? Can you see a fingerprint developing?*
- When you are done, gently blow off excess powder from the surface and check your result. *Did you get a visible fingerprint? How well can you see the print? Can you make out the fingerprint pattern?*

- Use a piece of clear tape and carefully press the tape down onto the surface on top of the developed fingerprint. Peel the tape off and transfer the fingerprint onto a piece of paper. (If you used dark cacao, use white paper; for baby powder, use black.) *Did you successfully transfer your fingerprint? Does it still look the same on the paper compared with your glass or metal surface?*

- Wash your hands thoroughly with warm water and soap. Then, repeat the steps with the same finger you chose before. *Did you get a fingerprint again? How does it look compared with the first one? Can you still see lots of details?*

- Finally, apply some hand lotion and repeat the fingerprint steps again. *How does this fingerprint compare with the previous ones? Is it easier or more difficult to detect?*

- Once you have transferred all your fingerprints onto a piece of paper, compare your results. *Do they all look the same? Which one can you see the best, the least? Can you explain your results?*

EXTRA

In this activity, you tested a nice, smooth glass or metal surface. *Do you think other surface textures or materials will result in fingerprints as well?* There is only one way to find out! Test other materials such as paper, textiles, or wood. *How do fingerprints look on these surfaces?*

EXTRA

Now that you are like a professional in collecting fingerprints from surfaces, try to find them in your house! *Where is the best place to look for them? Can you find your own or some from your family members and make them visible?* Be sure you have permission to dust on surfaces before you test them.

OBSERVATIONS AND RESULTS

Were you able to collect some of your own fingerprints? On a smooth surface such as glass or metal, fingerprints tend to stick very well. With your unwashed hands, you should have been able to make your fingerprint visible with either cacao or baby powder. Just a little powder applied with a brush should be enough to reveal your fingerprint. If you apply too much powder, however, the fine details of your print might have gotten lost. When you press too hard onto the surface with the brush, the fingerprint will be wiped away—so you have to be careful when treating the surface with the powder.

Your freshly washed hands have much less skin oil and sweat because they have been washed away with the soap and water. This results in a much less pronounced fingerprint. You might have had difficulties in collecting this fingerprint or may not have found one at all. On the other hand, if you apply hand lotion, which contains lots of oil and fat, this will make your fingertips much stickier, leading to a much more pronounced print. You should have seen a big fat fingerprint once you applied the powder to the surface where you had pressed your finger. If you compare all the prints you collected, the one with hand lotion should be most visible whereas the impression with your washed hands should be barely visible.

If you did the extra activities, you might have noticed porous or rough surfaces or materials such as paper or textiles are not very good for collecting fingerprints. Considering your results, where would you look for fingerprints in your home? Did you find some?

CLEANUP

Clean your brush, wipe your fingerprinted surfaces with the dust cloth, and dispose of the used baby or cacao powder in the trash.

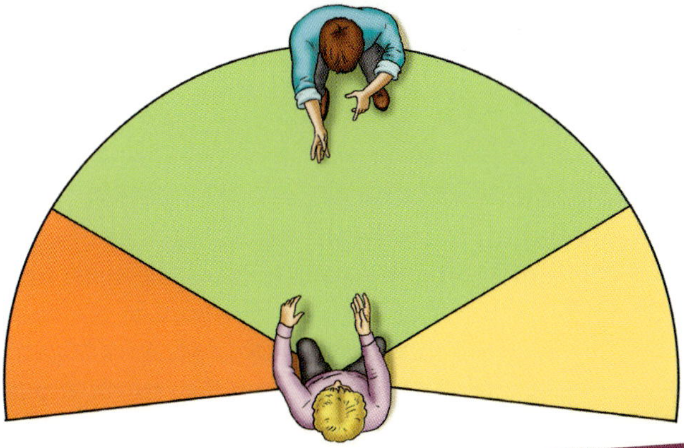

Exploring the Depths—of Vision!

NOW YOU SEE IT: HOW WELL CAN YOU CATCH A BALL WITH JUST ONE EYE OPEN? PUT YOUR DEPTH PERCEPTION TO THE TEST AND SEE HOW YOU DO!

PROJECT TIME

30–45 minutes

KEY CONCEPTS

Biology
Vision
Binocular vision
Depth perception

Did you know there is a type of giant clam that has thousands of small eyes covering its body? It's called *Tridacna gigas*, and it lives in the South Pacific Ocean. Although it has a lot of them, *T. gigas's* eyes are fairly primitive compared with ours. (So don't be too jealous that you don't have thousands of eyes too!)

All vertebrate animals (including humans) have two eyes these days. Humans have developed an amazingly complex visual processing system that allows us to collect and process a tremendous amount of information—in the blink of an eye! Our visual system is so good at processing information for us that we usually don't even notice what it's doing. In this activity, we're going to explore some of the amazing abilities of that system. So get ready to say, "Thanks, eyes!"

BACKGROUND

Having two eyes gives humans (and other animals) many advantages in our daily lives. For example, having two eyes allows us to see a larger field of view. (To test this, cover your left eye and try to see something on that side of your head. Without moving, you can't!) In addition, the eyes of humans and some other predators are special because they are positioned on the front of our heads. This gives us what is known as binocular vision, where our two eyes work together to see the world around us. Some animals, particularly prey like rabbits and mice, have eyes positioned on either side of their heads. These animals have what is known as monocular vision, where the information coming from each eye is different, and there is little to no overlap between the visual fields of the two eyes.

Having binocular vision gives humans and other animals much better depth perception, which is our ability to perceive our world in three dimensions and estimate the distance between objects. For example, when you reach across the table to pick up a glass, you use your depth perception to tell you where that glass is—and when to stop moving your hand toward it, so you don't knock it over.

In this activity, we'll be exploring exactly how much binocular vision contributes to our depth perception. Keep your eyes open!

MATERIALS

- Ping-pong ball
- Fabric, scarf, or eye patch (something to cover one eye)
- Partner
- Pencil or pen
- Paper
- Yardstick or measuring tape
- Removable tape (such as painter's tape)
- An open space, preferably indoors, of at least 10 by 5 feet (3 by 1.5 m)

PREPARATION

- Use your yardstick or measuring tape to measure a distance of 10 feet (3 m). Mark your starting point with a piece of removable tape. Mark every 2 feet (61 cm) with a piece of removable tape.

- At the 10-foot (or 3 m) marker, place a sheet of paper on the ground that says "10 FEET" (or "3 METERS") in large letters that can be read from your starting point.
- On your paper, make a table that includes rows for each of the increments (2 feet, 4 feet, etc.) and a row at the bottom labeled "Total (eye)." This column will be labeled "Distance." Add columns to the right: "Both eyes—successful catches (out of 10)"; "Right eye—successful catches (out of 10)"; "Left eye—successful catches (out of 10)"; and "Total (distance)."

PROCEDURE

- Stand at the starting point. Have your partner stand on the first marker, holding the ping-pong ball.
- Have your partner gently toss the ball to you 10 times. Ask your partner to observe you as you catch it. Notice how you track the ball with your eyes. *Do you watch it from the moment it leaves your partner's hands to the moment it lands in yours?*
- In the column "Both eyes—successful catches," record the number of times you caught the ball.
- Repeat these steps again, with both eyes open, at each distance you marked on the floor. Record your successful catches in the corresponding row. *Does your success rate change as the distance changes?*
- Use your makeshift eye patch to completely cover your left eye.
- Repeat the first step by standing at the starting point again. Have your partner stand on the first marker holding the ping-pong ball.
- Take a moment to notice how your vision is affected by covering one eye. Try reading the sign on the floor at the 10-foot (or 3 m) marker. *Is the sign more difficult to read? When you look around, do you see things less clearly?*

- Once again, have your partner gently toss the ball to you 10 times. Ask them to observe you as you catch it. *Is it easier or more difficult to catch the ball with one eye covered? Do you notice any changes in how you catch it? For example, do you keep your hands up higher or lower? Do you turn your head at all while you're tracking the movement of the ball? Does your partner notice you doing anything different to catch the ball?*

- In the column "Right eye—successful catches," record the number of times you caught the ball at a distance of 2 feet (61 cm).

- Repeat these steps with your left eye covered, at each distance you marked on the floor. Record your successful catches in the corresponding row. *Does your success rate change as the distance changes? Is the distance more challenging with only one eye open? If so, why do you think that might be the case?*

- Remove the eye patch from your left eye and use it to completely cover your right eye.

- Take a moment to notice how your vision is affected by covering your right eye. Try reading the sign on the floor at the last marker again. *Is the sign more difficult to read than it was when your left eye was covered? When you look around, do you see things less clearly?*

- Repeat the first step by standing at the starting point again. Have your partner stand on the first marker, holding the ball.

- Again, have your partner gently toss the ball to you 10 times. Ask them to observe you as you catch it. *Is trying to catch the ball different with only your left eye open compared with when only your right eye was open? Does your partner notice that you do anything different to catch the ball?*

- In the column "Left eye—successful catches," record the number of times you caught the ball.

- Repeat the steps with your right eye covered at each distance you marked on the floor. Record your successful catches in the corresponding row. *Does your success rate change as the distance changes? Is the distance more challenging with only one eye open? If so, why do you think that might be the case?*

- In the "Total (distance)" column, add up the total number of successful catches you had for each distance.

- Compare the "Total (distance)" results for each row. *At which distance did you have the most successful catches? At which did you have the least number of successful catches?*

- In your "Total (eye)" row, add up the total number of successful catches you had for each eye column.

- Compare the "Total (eye)" sums from each column. *Did you have more catches for "Both eyes" compared with "Right eye" or "Left eye"? Was there a difference between how many successful catches you had with your right eye open compared with your left?*

EXTRA

Repeat the procedure with your partner catching the ball. Pay attention to how they catch the ball when both of their eyes are open compared to just one. *Did they have different results with a different eye covered than you did?*

EXTRA

It's easy to test depth perception! Place a few items on a table in front of you, then close (or cover) one eye. Try to reach for the objects quickly, without trying to feel where they are. Make sure you don't use anything fragile!

EXTRA

Another way to test depth perception is to hold a pen or pencil in each hand. Moving just your hands, tap the points together. Try doing the same thing with one eye covered too.

OBSERVATIONS AND RESULTS

In this activity, you explored some of the functions of our binocular visual system. When you covered one eye, you probably noticed it was more difficult to catch the ping-pong ball compared with both eyes open. If you look at the "Total (eye)" row and compare the number of successful catches for both eyes open, right eye open, and left eye open, you probably can see that you were much better at catching the ball with both eyes open.

When you compare the right- and left-eye totals, you might have also noticed that you were better at catching the ball when just your right eye was open compared with your left—or vice versa. This is common and can be a sign of eye dominance. Just like most people have one dominant hand, many of us also have a dominant eye. This is the eye that our brain relies on most for visual information. Many people are slightly better at processing precise positional information (such as the trajectory of a ping-pong ball) that comes from their dominant eye. So if you notice a difference in the data you collected between your right and left eyes, you might have discovered which of your eyes is your dominant one.

In addition to eye dominance, this activity demonstrates the importance of binocular vision in depth perception. When you covered one eye, you probably didn't notice a big change in how clearly you could see things around you. You could probably still read things at a distance (as well as you could with both eyes open). Even though you could see just as clearly with one eye covered, however, you still struggled to catch the ball. This is because depth perception is improved when the brain receives information from both eyes. One reason for this improvement is the binocular visual cue known as stereopsis, or binocular retinal disparity. In short, having two eyes focused on an object allows us to triangulate its position with a higher degree of accuracy. Because of the distance between your right and left eyes, each perceives the object from a slightly different angle. If an object is far away, the disparity of that image falling on both retinas will be small. If the object is close, the disparity will be large.

To see this in action, hold your finger up in front of your eyes, about 6 to 8 inches (15 to 20 cm) from your face. Focus your eyes on something behind your finger, so you see a double image of your finger. Once you see two (slightly transparent) fingers, slowly move your hand away from your face. As you do this, the distance between the "two" fingers will appear to decrease. Cool!

Having two eyes allows us to do many things more easily, including judge the distances between objects. So next time you're playing a sport or running toward something or even reaching for your water glass—make sure to keep both eyes open!

CLEANUP

Put your measuring tool, pencil or pen, eye covering, and ping-pong ball away. Remove and dispose of tape and marker labels from surfaces. Thank your partner for helping out.

Ears
Do Their Design, Size, and Shape Matter?

ALL SHAPES AND SIZES: HOW DOES THE DESIGN OF ANIMAL EARS AFFECT HEARING? MAKE REPLICAS AND TEST THEM OUT ON YOUR OWN EARS TO SEE IF THEY GIVE YOU SUPER HEARING!

PROJECT TIME
30–45 minutes

KEY CONCEPTS
Senses
Sound waves
Hearing
Ear anatomy
Physics
Biology

Have you ever been puzzled by a faint noise nearby, trying to discover what it is? Maybe you turned your head or cupped your hand behind your ear, hoping to hear the sound better. What if we could make this cup huge? Some animals know the answer. Many animals with exceptional hearing have big ears. A serval (a type of African wildcat), for example, can hear a mouse wiggling its way underground. And bats, which rely on sound to help navigate in the dark, have notoriously large ears on their small heads.

In this activity, you will design and test your own earlike “hearing aids,” looking at animal ears for clues about what helps improve the auditory sense.

BACKGROUND

Sound travels through the air in the form of pressure waves. After these enter your ear canals (the holes on each side of your head), the inner ear (the part inside your head) jumps into action. It translates the sound waves into nerve signals—a sort of code—that are sent to the brain to process. It is only then that you hear the sound.

It seems like the inner ear does most of the work, so why do extensions of the ear that stick out on both sides of the head exist? It turns out these pieces act like funnels: They collect, amplify, and direct sound waves to the ear canal.

Pinnae—the scientific name for the extension of the ear that sticks out from the side of the head—are not randomly created. Take the human pinna, for example. Its twists and folds are such that they specifically enhance sounds with a pitch that is typical for a human voice, a sound humans care about. They enhance these sounds up to 100 times and leave other pitches untouched. In other words, it's a handy built-in listening tool that reduces background noise.

The human pinna also helps determine sound direction. Whereas sounds from the front and sides are enhanced by the pinna, those coming from the back are reduced. This leads to small differences in volume administered by our two ears. Together with the difference in arrival time, this helps us deduce the location of the sound source.

So now you understand a little more about the human pinnae. But what would it be like to have animal pinnae? Would immense earflaps (such as the ones elephants have) improve our hearing? Maybe we can try cupped pinnae that rotate? What about giant cones?

MATERIALS

- Heavy construction paper
- Scissors
- Tape
- A radio, CD player, or other musical device with speakers or a headset (Ear buds do not work well for this activity.)
- Two paper plates (optional)
- Markers or other decorative materials (optional)

PREPARATION

- To prepare, you will create two or three types of pinnae.
- First, create a pair of cone-shaped pinnae, also called ear trumpets. Roll a sheet of heavy construction paper in a wide cone. One side should have a hole that is small enough so it can rest in your outer ear, near the ear canal. (To avoid injury, do not insert anything inside your ear canal.) Use tape to secure the cone shape. Build a second identical cone to complete your pair of ear trumpets.
- For your second pair, create wide and flappy pinnae, like elephants have. Lay two pieces of heavy construction paper on top of each other and trim the edges to create two elephant earlike shapes.
- **Optional:** Create cupped ears by cutting a triangular piece from a paper plate. Imagine if your plate were a round cake; cut a quarter piece of the cake away and discard that piece. Holding or taping the two cut sides of the larger piece together creates a hole, making a nice cupped pinna to put on your ear. Make a second identical one for your other ear. Because you will put these pinnae around your ears, you might want to cover the cut edges with tape.
- **Optional:** Decorate the pinnae.

PROCEDURE

- Put the radio, CD player, or other musical device with speaker on low volume, so you barely can detect the sound. If you use a headset, leave it on the table and put the volume on high so you can hear a faint sound without wearing it. Stand close to the speaker or headset, with one ear turned toward it. Leave enough space so you can put an ear trumpet between your ear and the speaker or headset. Listen to the sound. *How does it sound? Faint, barely audible?*

- Place the ear trumpets in your ears so the small holes rest in the outer ear, close to the ear canal. Point the wide-open end of one cone toward the speaker or headset. Point the other cone in the opposite direction (away from the speaker or headset). Your head should be at approximately the same position as in the first step. Listen to the noise. *How does it sound this time? Is it slightly louder or fainter? Can you still hear the sound?* If it is louder, turn the volume slightly down until you can just barely hear the sound. Remove the ear trumpets and listen again. *Can you hear the sound from the same distance without the cones?*

- Put the ear trumpets back on so the small holes are pointing toward your ear canal and one of the wide openings points toward the speaker. Test to make sure you can still hear the sound. Now turn your head so you are looking at the speaker and the ear trumpets are pointing to the sides. *How does it sound this time? Are you still able to detect the sound?*

- To test what cupped pinnae might sound like, cup your hands around your pinnae, curving them forward a bit. Listen to the faint noise while you look at the speaker. *Does the hand help you hear the sound? Which hand position enhances the sound best?*

- Animals often tweak their ears. *Do you think tweaking your cupped pinnae will make a difference?* To test this, let the cup of your cupped hand point toward the sound source, then downward, and then backward (away from the sound source). *Does a particular position help you hear the faint sound better?*

- Some animals have huge, flappy ears. *Do you think these help them hear well?* To test this, place the wide and flappy pinnae—those that look a little like elephant ears—behind your outer ears so they extend your pinnae. Listen to the sound. *How does it sound? Do you need to adjust the volume to make it faintly detectable?* Turn your head to test all directions. *Did these actions help your hearing?*

EXTRA

Test your cupped paper-plate pinnae the same way you tested the hand-cupped pinnae.

EXTRA

To get a bit more creative, let's mix and match outer ears. *What happens if you combine an elephant-like pinna on the right side with an ear trumpet on the left? What happens if you cup your right hand facing forward around one ear and your left hand facing backward around the other?* Close your eyes and listen carefully. *How does the volume change? Can you also guess the sound's direction?*

EXTRA

Look at the shape, position, and direction of real-life animal ears in a book, on the internet, or as you come across animals in person. *How do you think the characteristics of their pinnae help that animal? Might these pinnae perform other tasks beyond hearing?*

EXTRA

Use your best design to go on a noise detection quest. *How many different sounds can you detect? Can you locate where they come from?* You might be surprised how many noises you can detect just by directing your attention and listening carefully.

OBSERVATIONS AND RESULTS

Did the ear trumpets and cupped pinnae improve your hearing when the wide parts or cups were pointed to the sound source but reduce your hearing when they were pointed elsewhere? This is to be expected, because pinnae serve as funnels for sound waves. Bigger funnels (like the ear trumpet or hand-cupping) collect more sound waves, so you hear the sound better. Because they are big, however, they can also hinder a sound from reaching your ear canal. That is why they made the sound appear fainter when you turned your head or cupped your hand backward (away from the sound source).

The big flappy elephant-like ears probably did not enhance your ability to pick up a sound. One of their non-hearing uses is to help animals cool down. These ears are also full of tiny blood vessels ready to release body heat. (Humans sweat to cool down.) Although large flappy ears are useful, they are not generally useful as hearing aids.

Unlike humans, many animals can deliberately move their ears (beyond a wiggle, that is). Some (such as horses) can even move each ear independently. They point the cupped ear, as needed, to enhance a sound and localize the source. Your brain is not trained to use ears that move, so when testing different directions, you might have felt a little confused, unable to identify the source well. A similar confusion can happen when you combine different types of pinnae. To effectively use these new pinnae, you would need to wear them for a period of time to retrain your brain.

The human pinna helps you focus on interesting sounds by selectively amplifying sounds with a pitch similar to that of a human voice. The pinnae you created are too simple to amplify specific pitches but animal ears or hearing aids can. As sounds get processed, your brain further helps you ignore background noise. Your noise detection quest will probably bring interesting sounds to your awareness, sounds you usually ignore. Your designer pinnae might still be very helpful to detect and localize faint sounds.

CLEANUP

Put away the tape, scissors, and the musical device. Dispose of the paper you used. Recycle if possible.

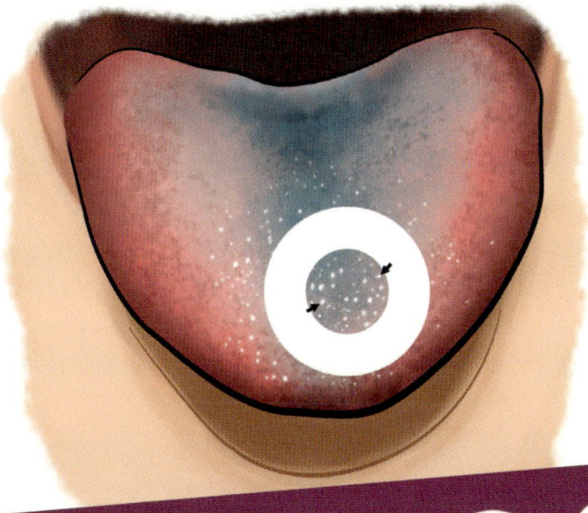

Super-Tasting Science

Find Out If You're a "Supertaster"!

TASTE TEST: TAKE A CLOSER LOOK AT THE TASTE BUDS ON THE HUMAN TONGUE.

Have you ever noticed that some people are a lot pickier about the food they eat than other people are? They might be more selective because they are supertasters! To supertasters, the flavors of foods are much stronger than to average tasters. Whether or not someone is a supertaster comes down to the taste buds on his or her tongue, and you can actually investigate a person's supertaster status by looking at this. Are you a supertaster? Find out with this tongue-based activity!

PROJECT TIME

30–45 minutes

KEY CONCEPTS

Human biology
Taste
Taste buds

BACKGROUND

Do you hate the taste of broccoli? Or think that grapefruit is extremely bitter? If so, you may be able to blame it on your taste buds! Taste buds, located on small bumps on the tongue called fungiform papillae, are each made up of about 50 to 150 taste receptor cells. On the surface of these cells are receptors that bind to small molecules related to flavor. Each receptor is best at sensing a single flavor: sweet, salty, bitter, sour, or umami. The sum total of these sensations is the "taste" of the food.

The number of taste buds varies from person to person. People who have relatively more taste buds are called supertasters. To supertasters, foods may have much stronger flavors, which often leads to supertasters having very strong likes and dislikes for different foods. Supertasters often report that foods like broccoli, cabbage, spinach, grapefruit, and coffee taste very bitter. The opposite of supertasters are non-tasters. Non-tasters have very few taste buds, and to them, most food may seem bland and unexciting. The people in the middle are average tasters. Which kind of taster do you think you are? What about your friends and family?

MATERIALS

- Paper-hole reinforcers (one for each volunteer)—or clean wax paper, a standard hole puncher, and scissors (instructions below)
- Water and soap
- At least five volunteers (such as friends and family)
- Blue food coloring
- One glass of water for each volunteer
- Magnifying glass
- A bright light or flashlight
- Paper and a pen or pencil

PREPARATION

- If you do not have paper-hole reinforcers, wash your hands with soap, take a clean piece of wax paper, and punch one hole in it for each volunteer. Cut out each hole so that you have at least five small wax paper "rings."

PROCEDURE

- Wash your hands thoroughly with soap.
- Have the first volunteer open his or her mouth. Place a drop of blue food coloring on the tip of his or her tongue. (Do not let the tip of the food coloring tube touch the volunteer's tongue.)
- Have the volunteer take a mouthful of water, swish it around in their mouth, and then spit it out.
- Ask the volunteer to make their tongue relatively dry by swallowing a couple of times. The remaining dye should stain all of the tongue blue except for the fungiform papillae, which will look like relatively large light blue or pink bumps (in a sea of dark blue). These papillae are where the taste buds are.
- Put a paper-hole reinforcer or a wax paper ring on the tip of the volunteer's tongue, in the area that has been dyed.
- Using a magnifying glass under bright light, count the number of papillae within the hole. *How many papillae are within the hole?* Write this number down along with the volunteer's name. **Tip: Don't count the really tiny bumps—just the larger ones.**
- Your volunteer can now remove the paper-hole reinforcer or wax paper ring from their tongue.
- Repeat the entire procedure with at least four more volunteers. Before testing each volunteer, wash your hands thoroughly with soap, and use a new cup with fresh water for each person. If you want to try this on yourself, have someone else count your papillae, or count them in a mirror.
- *How many papillae did most people have? Did people have different amounts of papillae, with some having a lot more than others? Knowing that if a person has more than 30 papillae they are considered a supertaster, were there any supertasters in your volunteer group? If so, how many?*

EXTRA

Repeat this activity with at least 25 more volunteers, for a total of at least 30 people. *What percentage of people in this larger group turns out to be supertasters? What percentage are non-tasters, having fewer than 15 papillae? What percentage has between 15 and 30 papillae and are average tasters?*

EXTRA

Try this activity again, but this time use an equal number of male and female volunteers, with at least 15 of each. *Do males and females have the same likelihood of being a supertaster?*

EXTRA

Look into how different foods, like spinach and chili peppers, taste to the different types of tasters. *Based on this, can you make a taste test to figure out who is a supertaster?* Check how accurate your taste test is by also counting the papillae for each person.

OBSERVATIONS AND RESULTS

Did you find that people had different amounts of papillae and that most people were not supertasters?

Typically when people do this activity, if they have more than about 30 fungiform papillae they are considered a supertaster; if they have around 15 to 30 papillae, they are an average taster; and if they have fewer than 15 papillae, they are a non-taster. Although the percentages vary around the world and with different populations of people, around 25 to 30 percent of people are thought to be supertasters, 40 to 50 percent average tasters, and 25 to 30 percent non-tasters. (If you tested only five people, you may not have seen that around 1 to 2 people, or about 25 to 30 percent, were supertasters because of the small sample size. Additionally, if you only used genetically related family members, this may have skewed the results as well since there is a genetic component to how many taste buds a person has.)

Testing a person's sensitivity to a bitter chemical called 6-n-propylthiouracil (PROP) is a more definitive way to determine if he or she is a supertaster. Non-tasters can't taste PROP, but supertasters can and really don't like its bitter taste! Average tasters can taste it too, but its bitter taste is not strong enough to bother them.

CLEANUP

Throw out any used paper holes or reinforcers, and put away the food coloring, magnifying glass, flashlight, and pen or pencil. Clean the water glass. Don't forget to thank your volunteers.

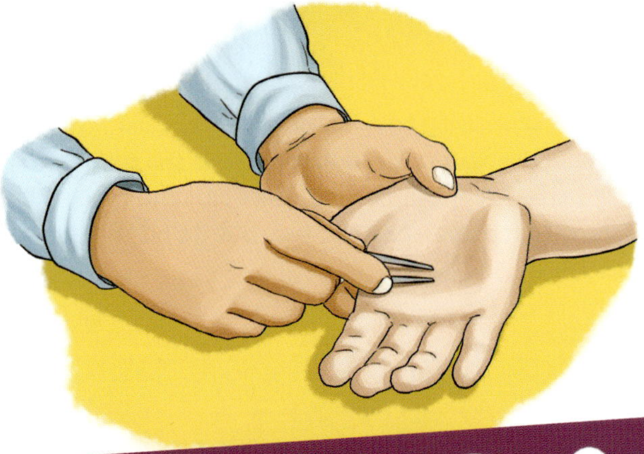

Skin-Deep Science

Find Your Sensitive Side

THAT TICKLES! DISCOVER HOW DIFFERENT BODY PARTS DIFFER IN SENSITIVITY.

Are you ticklish? If so, you've probably noticed that some parts of your body are more ticklish than others. That's because your skin's touch receptors aren't evenly distributed—some areas have more and others have a lot less. In this activity, you'll learn more about your sense of touch by testing your body's own reactions.

PROJECT TIME

30–45 minutes

KEY CONCEPTS

The five senses
The nervous system
Neuroscience

BACKGROUND

This activity is based on a physical exam used in medical checkups. Doctors use this test—called two-point discrimination—to study sensitivity on different parts of a patient's body. This sensitivity is pretty important—without it you might not notice when your finger gets too close to a flame or when you've scraped your knee. Thanks to your skin receptors, you can feel a soft fleecy blanket, a cold splash of water, or a sharp pinch—ouch!

Skin receptors are actually tiny cells called sensory neurons. When your skin makes contact with something—like a fluffy bunny—sensory neurons receive that "fluffy" signal and send out their own signals that travel through your body to your brain, which puts them all together to recognize that you are feeling a fluffy bunny.

MATERIALS

- Paper clips
- Ruler
- Pencil and paper
- A partner
- Glass with ice cubes (optional)

PREPARATION

- Bend a paper clip into a U-shape with the tips about 1/2 inch (1 cm) apart.
- Keep the glass with ice cubes in the freezer until the end of the activity.

PROCEDURE

- Tell your partner to close their eyes—no peeking until the end of the test!
- Explain that in this activity, you will touch different parts of your partner's hand and arm using a paper clip, and your partner will need to say whether they are feeling one point or two.

- Gently touch both tips of the paper clip to the back of your partner's hand. Both paper clip tips should touch the skin at the same time. *Do you think your partner can feel both ends of the paper clip?*

- Ask your partner if they feel one point or two. If your partner feels only one point, bend the paper clip U a little bit wider and repeat.

- Record the smallest distance between paper clip tips where your partner feels two points. The smaller the distance is, the more sensitive the area.

- Repeat each of the previous steps, this time touching your partner's palm; then try a fingertip, elbow, arm, and so forth. *Is your partner's sensitivity greater on the arm or elbow? Right or left hand? Palm or fingertip?* You can try this experiment on other parts of the body too—like the knee, back, toe, or cheek. *Why do you think some parts of the body are more sensitive than others?*

- Switch roles with your partner and find your skin's best-sensing spots. *Are your results the same? Is your palm more sensitive than your partner's? What about elbows?*

EXTRA

You can also try putting an ice cube on a spot like the back of your partner's hand for 10 seconds. Then use the paper clip to check whether that area is more or less sensitive. *Do you notice a difference? If so, why do you think there's been a change?*

OBSERVATIONS AND RESULTS

Were your fingertips more sensitive than your arm? Why do you think one area is more sensitive than another?

The part of your brain that receives information from your sensory neurons doesn't treat all parts of the body equally. The reason you are more sensitive on your fingertips than your elbow is that there are many more sensory neurons on your fingertips. When an area has more sensory neurons, there is a larger brain area devoted to receiving their signals, meaning more sensitivity. Most people find that their hands are much more sensitive than their backs or legs. Given how much you use your fingers for, that extra sensitivity makes good sense. You might even say it's handy!

CLEANUP

Try to bend the paper clip back into a usable shape. Put away the ruler, pencil, and the glass of ice cubes, if you used them. Thank your partner for their participation.

The Liver
Helping Enzymes Help You!

ENZYMES AT WORK: SEE HOW PROTEINS CALLED ENZYMES CAUSE CHEMICAL REACTIONS IN DIFFERENT SITUATIONS.

Your liver is important for cleaning up potentially dangerous substances you consume. How does it do it? Within your liver, as within every tissue in the body, many chemical reactions occur. Often these reactions require "help" to happen at a faster speed, and this can be supplied by enzymes—types of proteins.

The liver uses specialized enzymes to help it break down toxic substances and make them safer for the body to process. But an enzyme, just like the chemical reactions it modifies, needs certain conditions to do its work. So, some environments can make a liver enzyme effective, whereas others can prevent it from working at all.

PROJECT TIME

45-60 minutes

KEY CONCEPTS

Chemistry
Acids
Bases
Temperature
Physiology
Molecular biology

BACKGROUND

A chemical reaction occurs when compounds come together and their molecules interact to form new compounds. Sometimes these reactions happen by themselves, are usually fast and spontaneous, and give off energy. Other chemical reactions need energy, without which they would proceed slowly or not at all. Enzymes can help speed up these types of chemical reactions.

Enzymes are large proteins that speed up the rate of a chemical reaction by acting as a catalyst. A catalyst provides the necessary environment for the reaction to occur, thereby quickening it. Certain catalysts work for certain kinds of reactions; in other words, each enzyme has a particular type of reaction that it can activate. Enzymes can be very fussy and sometimes need to be in certain environments or conditions to work well—or at all. Some enzymes can even be damaged, such as when exposed to too much heat. A damaged enzyme may no longer work to catalyze a chemical reaction.

Catalase is an enzyme in the liver that breaks down harmful hydrogen peroxide into oxygen and water. When this reaction occurs, oxygen gas bubbles escape and create foam.

MATERIALS

- Raw liver (fresh or frozen, thawed; 1/4 pound [11.3 kg])
- Knife
- Cutting board
- Blender
- Water
- Refrigerator
- Medicine dropper
- Large plate
- Hydrogen peroxide (new or recently purchased bottle works best)
- Measuring teaspoon
- Two bowls
- Vinegar
- Baking soda
- Microwave-safe bowl (with a cover)
- Microwave oven

PREPARATION

- Completely disinfect any surface that the raw liver touches during this activity.

- On the cutting board, carefully cut the liver into little cube-shaped pieces, about 1/2 inch (1 cm) long. Be careful using the sharp knife. (An adult may need to help with this.)
- Place the liver cubes into a blender, and add an equal volume of water. Blend on high speed, pulsing when necessary, until the liver is smooth and no chunks are present. Be careful of the sharp blades in the blender.
- Keep the blended liver in the refrigerator.

PROCEDURE

- Put one drop of the blended liver on the large plate. Add one drop of hydrogen peroxide to the blended liver drop. You should see a lot of bubbles! *What do you think the bubbles are made of?* This shows that the liver enzyme catalase is working to start the chemical reaction that breaks down the hydrogen peroxide that would be harmful to the body into less dangerous compounds.
- To test the effect of an acid on the liver enzyme, put one teaspoon of the blended liver in a bowl and mix it well with one teaspoon vinegar. *What is the color and consistency of this mixture?* Put one drop of the mixture on a clean part of the large plate, and add one drop of hydrogen peroxide to it. *Compared with the untreated blended liver, did more, less, or about the same amount of bubbles form? Did they form more slowly, more quickly, or at about the same rate?*
- To test the effect of a base, put one teaspoon of the blended liver in a bowl and mix it with one teaspoon baking soda. *What is the color and consistency of this mixture?* Put one drop of the mixture on a clean part of the large plate, and add one drop of hydrogen peroxide to it. *Did more, less, or about the same amount of bubbles form? Did they form more slowly, more quickly, or at about the same rate?*

- To test the effect of heat, put one teaspoon of the blended liver into a microwave-safe bowl. Cover the bowl, and microwave it on high for 20 seconds. (An adult may need to help with this.) *How does the blended liver look after heating?* Remove a drop-size amount of the heated liver, and put it on a clean part of the large plate. Add one drop of hydrogen peroxide to it. *Did more, less, or about the same amount of bubbles form? Did they form more slowly, more quickly, or at about the same rate?*
- *Based on your observations, under which condition(s) does it look like the enzyme works best? Which condition(s) makes it work the worst? Why do you think this is so?*

EXTRA

Try experimenting with other conditions. For example, try freezing some blended liver or mixing it with salt and then test the enzyme's activity. Or you could try adding more than one teaspoon of vinegar or baking soda and then test the enzyme. *Under which conditions does the enzyme work well and under which ones does it work poorly?*

SCIENCE FAIR IDEA

You could try this activity again using another enzyme called bromelain, which digests proteins and can be extracted from pineapples. One protein that is fun to digest using bromelain is gelatin, which is found in many puddings and gelatinous desserts. *How do different conditions affect the ability of bromelain to digest proteins?*

OBSERVATIONS AND RESULTS

When exposed to hydrogen peroxide, did the blended liver bubble less when mixed with either the vinegar or baking soda compared with when it was untreated? Did it bubble even less after it was microwaved?

An enzyme needs certain conditions to work, and the ideal environment can be a hint as to where the enzyme normally works in the body. And because different body tissues have distinct environments, each enzyme is tuned to work best under specific conditions.

Different tissues in the body have different pHs (pH is a measure of how basic or acidic a solution is). The liver maintains a neutral pH (about pH 7), which is easiest for its enzymes, such as catalase, to work in. Consequently, when exposed to hydrogen peroxide, the liver should have produced more bubbles (oxygen gas), and at a faster rate, when it was untreated than when exposed to vinegar or baking soda. (It may have bubbled more when treated with baking soda, compared with vinegar, because it might have been better able to return the pH to around 7.)

Similarly, enzymes in the liver are also used to functioning at body temperature (around 98°F, or 37°C), so microwaving the blended liver to a temperature hotter than that should have damaged the catalase enzyme and clearly decreased the amount of bubbles when it was exposed to hydrogen peroxide.

CLEANUP

Safely dispose of any liver meat used in this activity by putting it in the trash when you are done. Completely disinfect any surfaces that the raw liver meat touched during this activity, and be sure to thoroughly wash your hands with soap and warm water. Put away any other materials that you used.

THE SCIENTIFIC METHOD

The scientific method helps scientists—and students—gather facts to prove whether an idea is true. Using this method, scientists come up with ideas and then test those ideas by observing facts and drawing conclusions. You can use the scientific method to develop and test your own ideas!

Question: What do you want to learn? What problem needs to be solved? Be as specific as possible.

Research: Learn more about your topic and refine your question.

Hypothesis: Form an educated guess about what you think will answer your question. This allows you to make a prediction you can test.

Experiment: Create a test to learn if your hypothesis is correct. Limit the number of variables, or elements of the experiment that could change.

Analysis: Record your observations about the progress and results of your experiment. Then analyze your data to understand what it means.

Conclusion: Review all your data. Did the results of the experiment match the prediction? If so, your hypothesis was correct. If not, your hypothesis may need to be changed.

GLOSSARY

amplify: To make something louder or stronger.
auditory: Having to do with the ears or hearing.
bronchi: Airways in the lower respiratory tract that bring air into the lungs. (One is called a bronchus.)
carotid: One of the two arteries in the neck that supply blood to the head.
corresponding: Matching something or having a direct relationship with it.
diaphragm: A thin, curved muscle below the lungs.
disparity: A noticeable difference between things.
distinguish: To see or hear something clearly.
dominance: The quality of being stronger than something else of the same type.
dynamic: The way that two or more things behave with each other because of a situation.
enhance: To improve or increase something.
forensic: Having to do with the use of scientific knowledge or methods in solving crimes.
gelatinous: Resembling gelatin, or jelly.
hinder: To make something hard to do or slow to happen.
notoriously: In a manner that is well known.
porous: Having small holes that allow air or water through.
replica: A close copy of something.
retina: The tissue at the back of the eye that receives images and sends signals to the brain.
sextillion: In the United States and Canada, a number equal to 1 followed by 21 zeros.
trajectory: The curved path along which something moves through the air.
triangulate: To find a location by measuring the distance between two known points and measuring the angles between each point and another, unknown point.

ADDITIONAL RESOURCES

Books

Biesty, Stephen, and Richard Platt. *Stephen Biesty's Incredible Body Cross-Sections.* New York, NY: DK Publishing, 2020.

Finan, Catherine C. *The Human Body.* Minneapolis, MN: Bearport Publishing Company, 2021.

Han, MeiLan K. *Breathing Lessons: A Doctor's Guide to Lung Health.* New York, NY: W.W. Norton & Company, 2022.

Lawton, Cassie M. *The Human Circulatory System.* New York, NY: Cavendish Square, 2021.

Websites

Discovery Education
sciencefaircentral.com

Exploratorium
www.exploratorium.edu/search/science%20fair%20projects

Science Buddies
www.sciencebuddies.org/science-fair-projects/project-ideas/list

Science Fun
www.sciencefun.org/?s=science+fair

Videos

Can I Eat That? What Is a Supertaster?
ny.pbslearningmedia.org/resource/nvsn6.sci.bio.taster/what-is-a-supertaster/

Hearing: How Your Ears Work
www.pbs.org/video/hearing-how-your-ears-work-diu18u/

Heart Structure and Function
ny.pbslearningmedia.org/resource/tdc02.sci.life.stru.circulator/from-the-heart/

INDEX